# SOLAR SYSTEM FOR KIDS: THE SUN AND MOON

Speedy Publishing LLC
40 E. Main St. #1156
Newark, DE 19711
www.speedypublishing.com

Copyright 2015

All Rights reserved. No part of this book may be reproduced or used in any way or form or by any means whether electronic or mechanical, this means that you cannot record or photocopy any material ideas or tips that are provided in this book

**The Sun and the Moon are the two objects in the Solar System that influence Earth the most.**

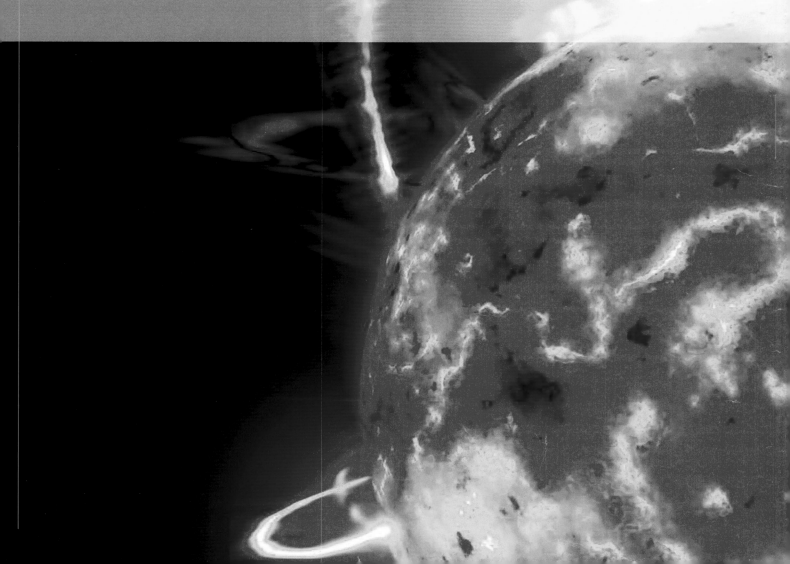

The Sun is a star found at the center of the Solar System. It is a nearly perfect spherical ball of hot plasma.

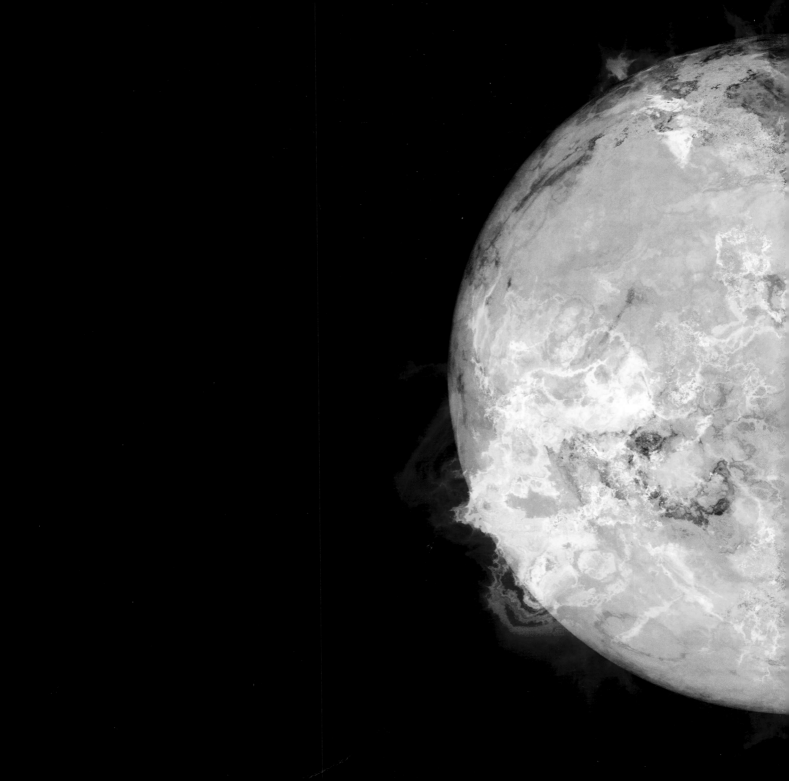

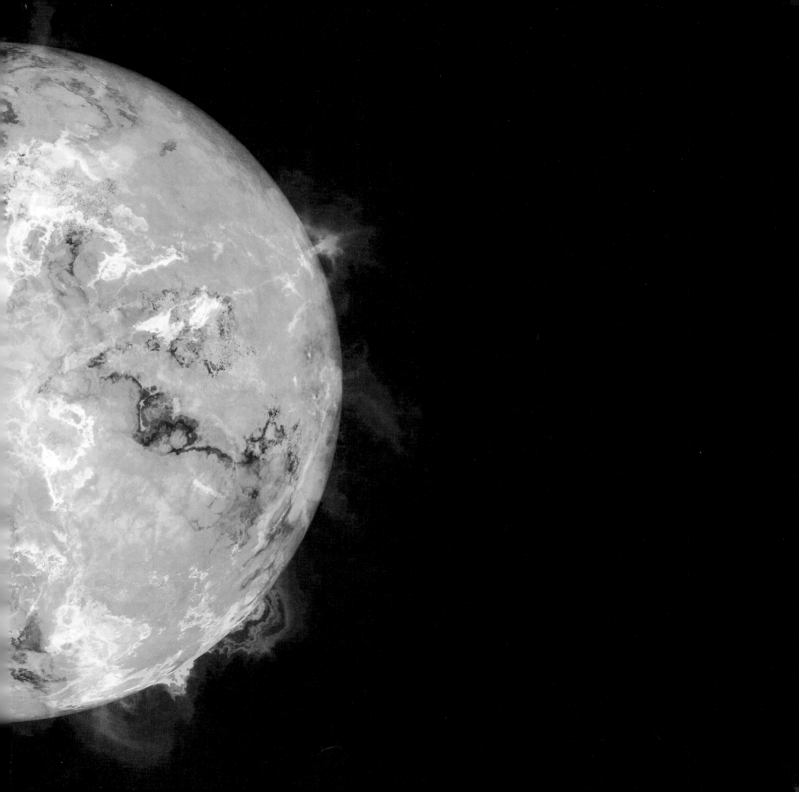

The Sun is over 4.5 billion years old. Its diameter is about 109 times that of Earth.

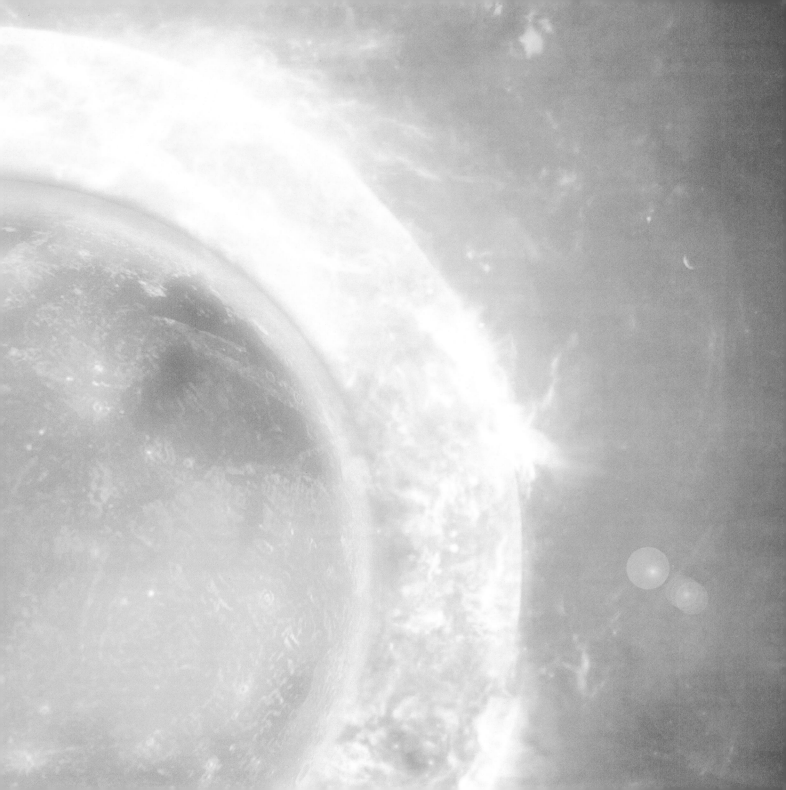

The Sun is the most important source of energy for life on Earth. The Sun controls our seasons. It controls how and when food grows.

It takes about 8 minutes for the sunlight to be seen on the earth after it has left the Sun.

The Moon is the Earth's only natural satellite. The moon takes about 29 days to orbit the Earth.

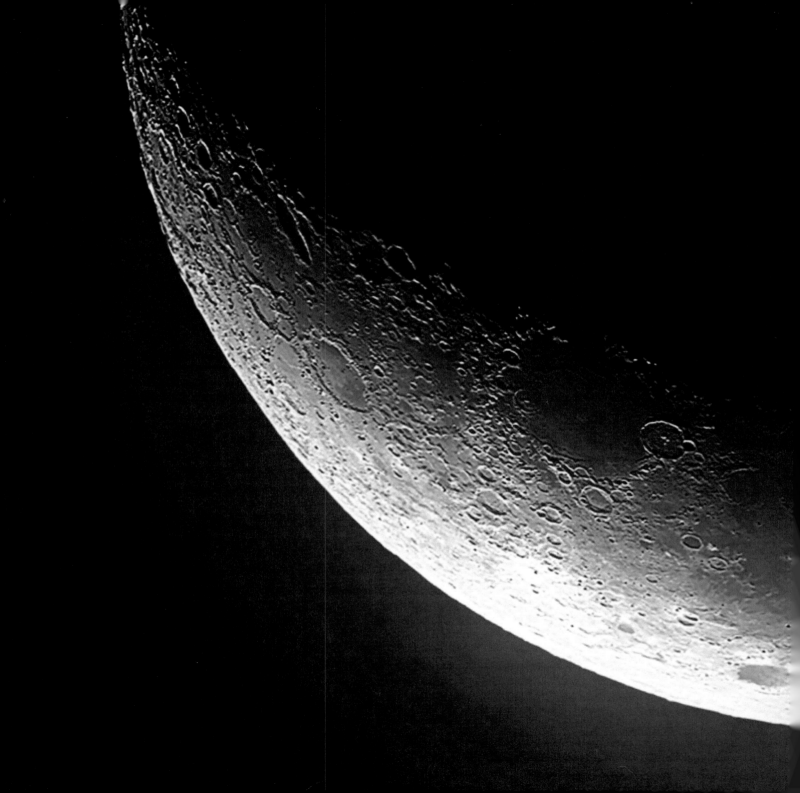

The moon causes many of the tides in the Earth's oceans. On the side of the Earth that is closest to the moon, the moon's gravity pulls the waters of the oceans up slightly, resulting in high tide.

The moon has no atmosphere. When the Sun is shining, the moon is 250 degrees Fahrenheit. When it is dark, the temperature reaches -280 degrees Fahrenheit.

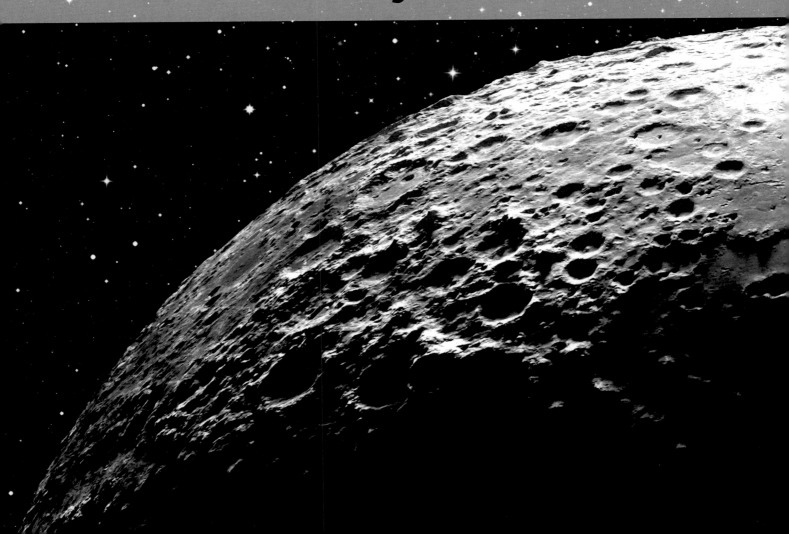

Made in the USA
Lexington, KY
01 November 2017